Stories Behind Theorems

Raffaella Mulas

Stories Behind Theorems

Conversations with Mathematicians

Raffaella Mulas ⓘ
Department of Mathematics
Vrije Universiteit Amsterdam
Amsterdam, Noord-Holland
The Netherlands

ISBN 978-3-031-96077-2   ISBN 978-3-031-96078-9   (eBook)
https://doi.org/10.1007/978-3-031-96078-9

© The Editor(s) (if applicable) and The Author(s), under exclusive license to Springer Nature Switzerland AG 2025

This work is subject to copyright. All rights are solely and exclusively licensed by the Publisher, whether the whole or part of the material is concerned, specifically the rights of translation, reprinting, reuse of illustrations, recitation, broadcasting, reproduction on microfilms or in any other physical way, and transmission or information storage and retrieval, electronic adaptation, computer software, or by similar or dissimilar methodology now known or hereafter developed.
The use of general descriptive names, registered names, trademarks, service marks, etc. in this publication does not imply, even in the absence of a specific statement, that such names are exempt from the relevant protective laws and regulations and therefore free for general use.
The publisher, the authors and the editors are safe to assume that the advice and information in this book are believed to be true and accurate at the date of publication. Neither the publisher nor the authors or the editors give a warranty, expressed or implied, with respect to the material contained herein or for any errors or omissions that may have been made. The publisher remains neutral with regard to jurisdictional claims in published maps and institutional affiliations.

Cover illustration: A glimpse into the magical office of Joost Hulshof. Photo by Raffaella Mulas, 2025

This Springer imprint is published by the registered company Springer Nature Switzerland AG
The registered company address is: Gewerbestrasse 11, 6330 Cham, Switzerland

If disposing of this product, please recycle the paper.

Preface

Mathematicians are often seen through a lens of stereotypes: logical, detached thinkers who live in a world of abstract concepts, numbers, and formulas. These perceptions can make mathematics—and mathematicians—seem distant, unapproachable, or even intimidating. But they only scratch the surface. Mathematicians are as diverse, vibrant, and human as anyone else, with personal stories, creative pursuits, struggles, and achievements that shape both their lives and their work.

This book brings together a selection of interviews that aim to go beyond the stereotypes and show the human side of mathematics. Each conversation offers a glimpse into the unique experiences, challenges, and passions of a different mathematician. These stories touch on the pursuit of knowledge, the balance between personal and professional life, reflections on mental health, and the central role of creativity in their work. Together, these voices show that mathematics is not only about logic and precision, but it is also shaped by diverse and often surprising perspectives.

Whether you are a specialist or simply curious about mathematics, stories, or both, I hope that these conversations bring you closer to the people behind mathematics.

Each chapter is based on a previously published article, and it has been a joy to re-imagine them here as part of a larger narrative. The chapters are based on the following articles:

1. **A conversation with Lisa Sauermann.**
   Published by European Women in Mathematics (EWM) and, in Italian, by MaddMaths!. To appear in the European Mathematical Society Magazine.
2. **Interview with Joost Hulshof.**
   Published on the website of the Mathematics Department of VU Amsterdam.
3. **Interview with László Lovász.**
   European Mathematical Society Magazine, 130:21–25 (2023).
4. **Proof by Example—Rianne de Heide.**
   Nieuw Archief voor Wiskunde, 5/24 nr.4 (2023).
5. **Interview with Jürgen Jost.**
   Published on the website of the Mathematics Department of VU Amsterdam.
6. **Interview with Cristiana De Filippis.**
   Published on the website of the Mathematics Department of VU Amsterdam. Submitted to the European Mathematical Society Magazine.
7. **Proof by Example—Renee Hoekzema.**
   Nieuw Archief voor Wiskunde, 5/25 nr.2 (2024).

I hope that you will enjoy "listening" to the words of these seven mathematicians as much as I did, and that their insights will spark curiosity and inspiration in you, too.

Warmly,
Raffaella Mulas

Amsterdam, The Netherlands   Raffaella Mulas

Acknowledgments

I am deeply grateful to Lisa Sauermann, Joost Hulshof, László Lovász, Rianne de Heide, Jürgen Jost, Cristiana De Filippis, and Renee Hoekzema, who generously shared their time, stories, and reflections with me.

These interviews were first published by the *European Mathematical Society Magazine*, *Nieuw Archief voor Wiskunde*, *European Women in Mathematics*, the website of the Mathematics Department of VU Amsterdam, and, in Italian, by *MaddMaths!*. I thank them for providing a space where these interviews could first reach the public.

I warmly thank the Max Planck Institute for Mathematics in the Sciences for allowing me to include their photographs of László Lovász and Lisa Sauermann. I am also honored to feature the beautiful artwork of painter Anneke Kerkhof and photographers Cressandra Thibodeaux and Annabel Jeuring that accompanies the interviews of Renee Hoekzema, Jürgen Jost, and Rianne de Heide, respectively. And a special thank-you goes to Anakin, who gracefully joins Cristiana De Filippis in her photo.

I am also grateful to my family, friends, colleagues, and the team of *MaddMaths!*, who supported and encouraged my interviews along the way.

I am grateful to the reviewers and to the Springer editorial team—especially Francesca Bonadei, Natasha Bailie, and Fairle T. Thattil—for believing in this book, helping it come to life, and patiently responding to my many emails.

And finally, I am grateful to you, for reading this book.

Contents

1 Breaking Stereotypes: Lisa Sauermann's
  Mathematical and Social Insights   1

2 Joost Hulshof's Enchanted World   7

3 Paul Erdős Legacy and Abel Prize Insights
  with László Lovász   13

4 Rianne de Heide: From Statistics to Music,
  with an Open Conversation on Mental Health   25

5 Science, Theater, and Intellectual Curiosity
  with Max Planck Institute Founder Jürgen Jost 33

6 EMS Prize Winner Cristiana De Filippis'
  Perspective on Research and Life's Balance   41

7 Renee Hoekzema: Blending Earth Sciences,
  Topology, and Artistic Expression   47

1

Breaking Stereotypes: Lisa Sauermann's Mathematical and Social Insights

Lisa Sauermann (Fig. 1.1) is a Full Professor at the Hausdorff Center for Mathematics (HCM) of the University of Bonn. She was born in 1992 in Dresden, Germany. She studied at the University of Bonn and she completed her PhD at Stanford University, before becoming Assistant Professor at the Massachusetts Institute of Technology (MIT) in 2021, and then Full Professor at the University of Bonn in 2023. Lisa is best known for her results in extremal and probabilistic combinatorics, as well as for her achievements at the International Mathematical Olympiad, where she won four gold medals and one silver medal, and where in 2011 she was the only participant who reached a perfect score.

In August 2022, Lisa visited the Max Planck Institute for Mathematics in the Sciences. She presented a work in which, together with other collaborators, she proved an old conjecture of Erdős and McKay, for which Paul Erdős offered $100. During this occasion, Raffaella Mulas took the chance to ask questions about herself and her career.

Fig. 1.1 Lisa Sauermann in Leipzig. *Photo: MPI MiS*

Thank you for taking the time to meet me, Lisa! I would like to start from the beginning, by asking you how you discovered your passion for mathematics, and in particular how old you were.

It was a gradual process, but it started when I was maybe ten or so. My parents knew that I liked riddles and puzzles, and one day my mother showed me the problems of some local math competitions for kids. I really enjoyed tackling the problems, so I decided to take part in these competitions. This is how I started to be exposed to the mathematics that they don't teach you in school, and I discovered that there was a kind of mathematics where you don't have to do computations with numbers. How was it for you?

It's funny because I can give the exact same answer you gave, with the only difference being that in my case it was my father who showed me the problems of the local math competitions. In fact, it might have been the same year as when it happened for you, since we are the same age.

And what was it about these competitions that interested you the most?

I liked the math problems, but I also really enjoyed the social aspect very much. It was nice to take part in the competitions and to meet other kids who had my same interests. And at some point, maybe when I was around 12, I started to travel to other cities and stay overnight, without my parents, to take part in national competitions. I had a lot of fun. I got to know many interesting people and new friends, and I got more and more fascinated by the social component of these events.

Interesting, so it seems that the social aspect had a big impact on your choice of becoming a mathematician.

Yes, it definitely had a very big impact.

How did your perception of mathematics change after you started to do "real math" at university?

It changed completely! When you take part in competitions, you have four hours for solving a set of problems, and you know that the solution must be relatively short. Research is completely different, because you could think about a problem for months, or even years.

And what is one thing that didn't change?

One thing that didn't change? This is a good question. Let me think about it... Maybe, the fun I have! This didn't change.

Well, this is a good answer! Now, your mathematical work is mainly in the area of combinatorics; how did you end up working on this topic?

Typically, people who take part in math competitions end up taking many combinatorics courses and working directly on this area, because it is usually the focus of such competitions. But I think it's good to explore different areas of mathematics, and when I did my undergrad in Bonn I didn't do any combinatorics. In fact, I focused on algebraic geometry, and I wrote my bachelor thesis in 2014 with

Michael Rapoport. It was only when I was in Stanford for my PhD program that I changed my mind and I started to focus on combinatorics instead. This happened after following Jacob Fox's course on extremal combinatorics. I first took this course just for fun, but then I liked the class so much that I decided to work on these topics with him. And I'm very happy that I did this!

Funny, I just realized that we almost overlapped in Bonn. The mathematical world is so small sometimes! And who have been your greatest supporters throughout your career?

My parents have always been supporting me so much. In fact, I couldn't be here talking to you right now if it wasn't for them, as they are currently looking after my kids. Of course my husband, Laurent, who is also a mathematician, as well as my PhD supervisor, Jacob Fox.

Can you tell me about a key moment where your PhD supervisor supported you?

During my PhD, after proving my first results, I had one year when nothing seemed to work, and this was very hard for me. Now I think that many grad students or also more senior mathematicians go through phases where things just don't work. But whenever you are in such a phase, you always think that all other people around you are succeeding. In particular, on arXiv you only see the successes, and there is no arXiv of things that people tried and didn't work. So whenever you are in this situation, you feel like you are the only one who is living it, and this can be very discouraging. When I was in this phase, Jacob supported me very much. He showed me how much he believed in me. Knowing that he thought I could do it, helped me to build up my confidence again, and taught me not to give up.

I find it beautiful that you are sharing this. Another thing that I found very nice earlier was that you said

that the social aspect of the math competitions had a big impact on the choice of your career. This goes completely against the stereotype that mathematicians don't have social skills. Are there any other stereotypes that would you like to break?

Yes, the list is so long! For instance, that mathematicians calculate with large numbers every day, or that mathematicians are all weird and unsocial.

Yeah, I hate these stereotypes as well. And how about the stereotype that women cannot become mathematicians?

I think that this stereotype is highly cultural. I grew up in eastern Germany, and for some reason, I never had the perception that society thought that women could not become mathematicians or scientists. But now that my daughters are growing up in the US, I see more stereotypes. Let me tell you two stories about my 3-year old daughter. She likes to play games that have to do with construction. One day her backhoe was broken and it needed to be fixed, so I gave her a female character from a Lego Duplo set, and said: "Maybe she can repair it!". But my daughter said: "No, women cannot fix backhoes!". I was shocked, because this idea certainly didn't come from me or my husband. And this was a negative story, but the next one is a positive one and of a similar nature. Another day, while playing with the same Lego Duplo set, she was looking for a character who could drive the backhoe, and again, I suggested for her to use a female character. She said: "No, women cannot drive backhoes! Women drive trucks!". And the reason she said this is that she once saw a female truck driver on the street in front of our house. I found it fascinating that seeing a female truck driver once completely changed her perception of what women can do. So, because children absorb a lot by observing, my guess would be that if they see even just a

Fig. 1.2 Lisa Sauermann (left) and Raffaella Mulas (right) at the Max Planck Institute for Mathematics in the Sciences, September 2022

few women in mathematics, then this could make a huge difference.

I completely agree with you! And you are giving a beautiful message with these stories. Thank you for sharing all of this. Can I take a selfie with you? (Fig. 1.2).

2

Joost Hulshof's Enchanted World

Joost Hulshof (Fig. 2.1) is a Professor of Mathematics at VU Amsterdam.

Raffaella Mulas interviewed him in September 2023.

Thank you for welcoming me in this wonderful office. You know, when I still didn't know you and you were on leave, I could see from outside your office that you had photos of the Beatles everywhere and various vinyl records, including a vinyl by Demis Roussos. I thought: "I cannot wait to meet this person!". I asked someone: "How will I recognize Joost Hulshof?" and they told me: "When you'll see him, you will know it's him!". This turned out to be true. And I'm very happy to have the chance to interview you now. Can you tell me something about your life and career?

I was born in 1959 in The Hague. In high school I liked mathematics and chemistry, and in the end, I chose mathematics. I studied in Leiden, and I also did my PhD there. Then I went to Minnesota for a postdoc, and I came back to the Netherlands to do another postdoc in Delft.

Fig. 2.1 Joost Hulshof in his office. *Photo: Raffaella Mulas*

In 1988, I became an Assistant Professor in Leiden, and I stayed there until 2000 or something, when I moved here as a Full Professor of Analysis. Then, I told myself that I was only going to hire people who are smarter than me. I was very happy when Rob Vandervorst and Jan Bouwe van den Berg joined the department, as I have learned a lot from them!

I really like your hiring philosophy! Can you also tell me something about the very interesting objects that surround you in your office?

Well, when the department had to move from the old building to this one, it was announced that we were only allowed to bring one box with things with us, and nothing else. But I took six blackboards from the old building and convinced them to take one of them here. Now there are more. I also have many books, some neckties—my favorite one is the one with Donald Duck, and nice furniture that I bought at a recycle shop. The cabinet next to you is ancient! Oh, and I was able to put a doll outside the window of my office, but I have already shown it to you, right?

Yes! The doll is amazing.

Then this is a copy of the Declaration of Independence, which was printed in the original fashion 200 years later. You know, sometimes my wife says that at home we need to clean up a little bit, and when I don't want to throw things away, I bring them to my office.

And why do you have an open umbrella?

I bought it at the Panorama Mesdag Museum, in The Hague. Mesdag was a great Dutch painter, and Panorama Mesdag is a cylindrical painting that he realized. It's really impressive, and it's huge. You can see a copy of the painting inside the umbrella! I used the umbrella only once, and then I decided to put it up there. Then, under the umbrella there is a pillow that I had to move from our couch because we have too many pillows, and under the pillow there is a wheel. I borrowed it from the physics department, where they used it for some outreach activity. On the pillow there is also a tube that one can use to shoot paper balls, and a drumstick from the last rock concert I went to just before the Corona pandemic. The drummer threw it to the audience at the end! Then, this is a doll that my sister made, and that's a bell that I bought in France. We use it for the Mathematics Colloquium. I also have a bottle of whiskey that Bob Planqué gave me with the idea of offering it to everyone whenever we want to celebrate new results—it's good whiskey!

Very nice! I have heard that you sometimes bring a blackboard with you when you go on vacation. Is this true?

Yes, that happened twice. It's this blackboard here.

Fantastic! How do your mathematical ideas take shape?

If I'm interested in a problem, I try to tackle it using different techniques. I also talk a lot to other people, in order to get inspiration, and to see different points of view. I always try not to be afraid, which is very important, and if a problem is too hard, then I take a break and I come back to it later on, after learning new things that might be helpful. I always keep a list of problems that I would like to return to at some point.

Your research spans so many different areas of mathematics. What are you working on now?

Now I'm mainly working on a system of nonlinear reaction-(degenerate) diffusion equations which arises in population dynamics. Linearising them, we ended up with very specific linear operators, and we need to understand their spectral properties first.

I see that you are using A3-papers to write mathematics–I do this when I visit my dad, because he is an artist and he only has big sketch paper around the house. What's your motivation?

I just like to have more room for writing. Writing on A3-paper is a bit like writing on the board!

What advice would you give young researchers?

This is a difficult question because today, for young mathematicians, life is much harder than how it was for me. For me, it wasn't too difficult to get the positions I got. My general advice would be, create a good network of collaborators, and don't be afraid when you do research!

This is good advice! Besides mathematics and music, what makes you happy?

I enjoy having fun! I like going out with friends and socializing. And spending time with my wife makes me happy! We share the passion for music, and we do many things together.

You love Rubik's cubes as well, right? What's the most interesting one that you have?

Yes, right! My favorite one is this one, but actually it's not a Rubik's cube. It's called Megaminx, and it has 12 faces. The Rubik's Cube came on the market when I was a student. Look, there's a little booklet here. It's in Dutch, and as you can see it had the price of 7.50 Dutch Guilders. It was written in 1981 by Jan van der Craats. Here it is written "For Joost Hulshof", by the author!

Wow! Your office is like Mary Poppins' magic bag, where anything can appear!

Haha, yes! When I read this booklet, I got fascinated. It has some serious algebra in it, and it really made me love algebra. And for my PhD thesis, I also proved some results on the Rubik's cube.

This is amazing! Thank you so much for sharing all of this. May I also take a picture of you in your amazing office, for the article?

3

Paul Erdős Legacy and Abel Prize Insights with László Lovász

László Lovász (Fig. 3.1) is a Hungarian mathematician and a Professor Emeritus at the Eötvös Loránd University in Budapest. He was awarded the 1979 SIAM Pólya Prize, the 1982 and the 2012 Fulkerson Prize, the 1999 Wolf Prize, the 1999 Knuth Prize, the 2001 Gödel Prize, the 2006 John von Neumann Theory Prize, the 2007 János Bolyai Creative Prize, the 2008 Széchenyi Prize, the 2010 Kyoto Prize and, most remarkably, the 2021 Abel Prize, which many consider to be the Nobel Prize of Mathematics. He is the former President of the International Mathematical Union, and the former President of the Hungarian Academy of Sciences. He was also one of the main collaborators of Paul Erdős.

Raffaella Mulas (Fig. 3.2) interviewed him in June 2023, while visiting the Alfréd Rényi Institute of Mathematics in Budapest.

Thank you so much for taking the time to meet me. It is an incredible honor and a privilege for me to interview one of the mathematicians I admire the most. I would like to start from the beginning. As a teenager,

Fig. 3.1 László Lovász in Leipzig, in November 2022. *Photo: MPI MiS*

Fig. 3.2 László Lovász and Raffaella Mulas at the Max Planck Institute for Mathematics in the Sciences, in November 2022

you earned three gold medals at the International Math Olympiad. You also won a Hungarian TV show in which students were placed in glass boxes and asked to solve mathematics problems. Is this true?

Yes!

Is this when your passion for mathematics started, and what drove you into mathematics at such a young age?

Well, it started a little earlier, maybe in the 8th grade, when I joined the math club of my elementary school. I really enjoyed working on the problems that were posed there, and the teacher of the math club, who was also the director of the elementary school, recommended us to subscribe to a Hungarian journal of mathematics for high school students. The journal was established in 1893, and I think it's the oldest one in the world which is still functioning. And that was a great experience! Paul Erdős used to write for the journal as well. He liked to pose some open problems that were easy to formulate but difficult to solve, and he always presented them together with some historical remarks. This was really very inspiring!

So, you were still in elementary school when you first read something written by Erdős for this journal, right?

Yes, I think so! One of the first issues that I was looking at had an article of Erdős about combinatorial geometry. Anyway, the teacher of the math club also recommended that I apply for the Fazekas Mihály Gimnázium, a high school that was starting a specialized class for mathematics. The Fazekas Mihály Gimnázium then became quite famous precisely for its mathematics classes, but it also attracted other good students in other areas. For instance, while I was there, in a parallel class there was Éva Kondorosi: she is a biologist and one of the Chief Scientific Advisors of the European Commission. So, it is a very good high school, in general.

There I met several other young people who were recruited for the same class, and that turned out to be an excellent community in which to learn and do mathematics. The four years I spent there were really fantastic in my life! And since the mathematics class was newly established when I joined, mathematicians from both the university and this institute [Alfréd Rényi Institute of Mathematics] were very interested in what was happening there. They used to give some afternoon classes at our school, and some of us started to regularly visit some of the professors, from whom we got new theory to read or problems to think about. So, many of us started doing some research during high school.

Wow! Well, this all worked out very well! Let's jump ahead now: Your work spans many areas of mathematics and theoretical computer science. What has been the most exciting research project for you, so far, in your career?

Oh, I think this has changed over time. Looking back, probably my largest project has been graph limit theory, which started in the early 2000s. When I was at Microsoft, several of us started to work on it, including my wife Kati [Katalin Vesztergombi], Balázs Szegedy, Vera Sós—who passed away a few months ago, very sadly—Jennifer Chayes, Christian Borgs and Lex Schrijver, whom you might have met or will meet in Amsterdam. And others have played some role or contributed at various stages too. I think that this has been my largest project. I have always liked things which connect different areas of mathematics, and the name "graph limits" already indicates that there is a connection going on between graphs and limits.

But I also liked, mostly in the 70s and 80s, when the theory of computing was developing. To me, it was clear that this was a mathematical theory, and that was a very exciting period. I am not sure I contributed so much to that, but I was interested, and I wrote papers. What was exciting as well is that it led to some new graph theory problems. I worked on them also together with Tibor Gallai, who was my mentor during university. There was no official PhD supervisor at the time, but he put in a lot of time and energy to help me to get ahead. I remember that he said, "Look at these two problems: the Hamilton cycle problem, and the matching problem. The matching problem has been solved in almost every possible sense, and the Hamilton cycle is very similar. Why is it so difficult, then?". And so, many of us started to think that maybe there was a reason for that. We started to think about it in terms of computational complexity, but we didn't get the right approach there.

We then tried to work on Kolmogorov complexity, to see if there is any difference, but that also didn't work out.

Then, in 1972–1973, I did what we would now call a postdoc. I went to the States, to Vanderbilt University, for one year, while my friend Peter Gacs, who was also interested in this topic, went to Moscow. There, he worked with Kolmogorov and with Leonid Levin, who was a student of Kolmogorov and who developed the P and NP theory essentially in the same way as (in the west), Stephen Cook and Richard Karp developed it. So, Peter Gacs and I both spent a year abroad, and when we met again, we immediately told each other that we could finally see the difference between the matching problem and the Hamilton cycle problem. We were very enthusiastic! And after that, for two weeks we even thought that we could prove that P is not equal to NP. Our proof was nice, but at the end it wasn't right, as it proved something weaker. But anyway, we kept focusing on this area, and we organized a seminar here [at the Alfréd Rényi Institute of Mathematics] and we kept trying to see how computational complexity could be handled mathematically.

Amazing! What is your creative process? How do your mathematical ideas take shape?

Well, I mean, it's of course always a back and forth between trying to solve a problem and then trying to apply the ideas. Maybe one thing which I like probably a little bit more than most of my colleagues is to, sort of, clean up a proof. I don't like to write it down until I get the most essential part of it. And that sometimes is useful because it can lead to a better understanding of the situation. I'll give you one example: I was still in high school, and I was not satisfied with the fact that graph theory was sort of very elementary, and therefore looked down upon by many mathematicians. So, I thought that there should be some kind of algebraic side to it. I reinvented how to multiply two graphs with a new type of strong multiplication, and I thought, "Okay, it's easy to check that this product is commutative and associative; but do we have the cancellation law? Does $A \times C$ being isomorphic to $B \times C$ imply that A is isomorphic to B?". I began to think about it, and eventually, around the end of high school, I came up with a proof. But it was quite complicated, so I wasn't satisfied with it. And I still remember when I realized that, if I don't count subgraphs, but instead I count homomorphisms in the proof, then the claim follows immediately. So, this reinforced my idea that you have to understand what moves the proof, not only to come up with the proof. And I think that is something I like to do all the way. If I prove something, I try to understand what is the best way of looking at it.

This is great advice! Why is Budapest the hometown of most of the greatest combinatorialists and discrete mathematicians in history, including yourself? Is there something in the water here?

Haha! Well, there are various explanations. For one thing, which I think is very important, I have to go back a long time. So, Hungary had struck a deal with Austria in 1867 to obtain a certain degree of independence. There was a liberal government in Hungary, and they did many important things, and two of these are relevant. One is that they established general public education for everybody. The other one is that they gave equal rights to Jews. So, there was a large Jewish immigration to Hungary around the end of the nineteenth century and early twentieth century, and the Jewish people sort of created a city life and a scientific life. I'm not saying that Hungary was unprepared: There were already first-class scientists in Hungary, including János Bolyai in mathematics. But anyway, all of a sudden, this mathematical life began to take place, and this is when, for example, this high school mathematics journal which I mentioned was established.

The Hungarian National Mathematical Olympiad was also established around the same time, in the 1890s. So, many talented young people were discovered, and this gave an important push.

Now, why discrete mathematics and graph theory? It started with Dénes Kőnig, whose father was also a mathematician; but the Kőnig's Theorem in graph theory is called after the son. Some version of this theorem came out of Frobenius' study of determinants. There is a famous Perron-Frobenius theory about non-negative matrices, and Frobenius was interested in knowing, for a matrix whose entries are variables and some of them are zero, whether the determinant is an irreducible polynomial of the variables. Then, Kőnig wrote a paper where he basically showed that this is all just a combinatorial problem about seeing which variable goes where, and he used bipartite graphs to illustrate the arguments. What's interesting is that he didn't prove "the" Kőnig Theorem (which in this special case amounts to characterizing when the determinant is identically zero), but he just reformulated Frobenius' proof using graphs. Frobenius then wrote another paper in which he did not say very nice things about Kőnig, as he was very much against translating the problem into graph theory; although this is one example where you have to get rid of all the unnecessary signs, sums and everything, and it's all just about the perfect matchings. So anyway, Kőnig got interested in this and then he wrote a textbook in 1937, and he had at least two students, Paul Erdős and my advisor Tibor Gallai. And so, they moved the theory ahead, and many other Hungarian mathematicians got interested in graph theory as a result.

Well, just in case, I will bring a tank of water from Budapest with me back to Amsterdam! Now, you have mentioned Paul Erdős several times already, and you have been one of his main collaborators: how would you describe him?

He was a very unusual person. Unlike the general picture often painted of him, he was very much concerned about other people. He knew about everybody, what they were doing, and he helped whenever anybody needed either a little money, some recommendation letter, or anything like that. But he didn't care so much for himself.

He couldn't visit Hungary during the Stalinist times, so it was only maybe near the end of the 50s, when he came back to Hungary for the first time after the war or after the Stalinist regime ended. I remember when I was young, maybe a young university student or maybe even a high school student, he was staying in a hotel when visiting Budapest. He was sitting all day in the lobby of the hotel, surrounded by young people, who were between 18 and 35 years old, or something like that, and he was sort of simultaneously working with several people on different problems. "Do you have any idea how to solve this? Oh, I have this additional question, maybe that's easier, or maybe that's also interesting." And sometimes, at lunchtime, he invited whoever was there for lunch at a restaurant. This was very inspiring, and I learned a number of things from him; not only mathematically, of course, but also on the human side.

He always thought that mathematics should be done publicly. He thought that, if you have an idea or have a new result, you shouldn't be afraid of sharing it with other people, because if they contribute to it or carry it on, then it will just be a better result—so you shouldn't try to keep it to yourself! He was always very unselfish, and on at least two occasions, when I was young, he gave me credit which wasn't unjustified, but it was maybe more than I deserved. The first case was when I first met Erdős, and he was already working with Lajos Pósa; you probably know the name. They had almost finished a paper. Pósa was a classmate of mine, and a good friend, and he asked me whether I could prove one of the results in the paper. So, I thought about it, and after a couple of days, I was able to prove their claim. Now, of course, if you know that something is true, then it's much easier to prove it. But anyway, Erdős added a footnote in their paper saying that this result was independently proved by László Lovász, which is not quite true.

The other case was (what is still called) the Lovász Local Lemma, which appeared in a joint paper of ours. Erdős emphasized that this particular lemma was mine, as he realized that it had broader implications compared to the rest of the paper, so he called the lemma after me. But it appeared in a joint paper, so, according to the standard rules, the lemma should be called Erdős-Lovász Lemma, if a name is needed at all. So, he was a very interesting person!

And from what you are describing, it seems like he was also very generous, both as a person and as a mathematician.

He was very generous, yes!

And how old were you when you first met him?

I was in high school, and I think my friend, Lajos Pósa, introduced us. I don't remember actually where it happened: probably during one visit of Erdős to my school. He used to visit the school about once a year to give a talk for the students there.

So, he was a very active mentor for young people as well.

Yes!

Do you have any other particular memory about Erdős that you would like to share?

There were several occasions when either we were in the US, or he visited Budapest and stayed with us for a week or two. This was often a little strenuous because he slept very little, and he liked to work for the rest of the day. But of course, we had teaching duties, and kids, and everyday life, and he understood that—but it was clear that he was rather impatient, and he wanted to sit down with us and work as much as possible. He was much older than we were, so it was really a little bit embarrassing to say, "Sorry, we are already tired!"

Does an Abel Prize laureate ever have difficulties?

Well, I'm teaching a class, and many more students come than before, because they want to see what an Abel Prize laureate looks like. And of course, just like everyone else, Abel Prize winners also suffer from bureaucratical difficulties: do these count as difficulties?

Yes, definitely! And when proving theorems, do you still sometimes feel stuck?

Of course! In mathematics, 90% of the time you are on the wrong path. I mean, you can't see the end of the path, so you have to try! It's also important to be able to turn back.

You mentioned your wife Katalin Vesztergombi, to whom you dedicate all of your books, and if I'm not wrong you also have many children and grandchildren...

Yes, we have four children, if this counts as many, and we have seven grandchildren, so far. But our son just got married half a year ago, so we still hope to have even more!

Besides mathematics, your large graphs, and your (large) family, what makes you happy?

There is definitely nothing comparable to these! But I like to walk in the nature and just look around, and this is one thing which I try to do regularly with my wife.

This is nice! What are you looking forward to in the future?

Well, if you are 75, you have to realize that, mathematically, you cannot really expect to have a career change. But there are some areas that I'm interested in seeing if they lead anywhere, mostly in and around graph limit theory. I am interested in trying to develop limit theory for graphs that are neither very sparse nor very dense, but are sort of in the middle range. There are some papers, but it's still rather far from being able to call it a theory, or something. I've also been thinking about developing some limit theory for matroids, or at least to generalize some of this matroid theory to some kind of continuous rank function, in the spirit of John von Neumann's continuous geometry. So yeah, there are interesting questions in all areas, but at this moment, I am looking at these two areas seriously. Now I also have more time to do mathematics, and there are some very good people with whom I work here, so I really enjoy this.

Well, I'm looking forward to reading your future results! What advice would you give young mathematicians?

During my university years, I have always found it very good to get interested in all areas of mathematics, and not only in my research area. I think that this proved to be very useful in my life, in my research. My advice is to not specialize too early, if possible. But I also understand that our current system is different now, and students have to specialize earlier. Part of this simply comes from the fact that the subjects are growing, so you inevitably end up in one branch which is already difficult enough to learn. For instance, when I was young, graph theory consisted of one or two books, and much of it, if you read it today, would be considered to be very elementary. And other areas are, of course, growing just as fast. So, it's difficult, but I think it's good still, to have some idea of what kind of goals other areas have. What is the main thing that they try to say? What kind of goals do they have? What's the advantage of looking at it in one way and not the other? There are big areas of mathematics where I have very, very little idea about what's going on, but I try to understand a little bit of that, nevertheless.

I like this advice. My last question is, why do you think it is important to keep studying graphs today?

Erdős had this idea that, if there are questions, you have to ask them. And, especially in graph theory, he was great at finding good questions—questions that lead to other questions. Eventually, this led to many branches of graph theory that were developed based on his questions and conjectures. Nowadays, the use of large graphs and large networks in various parts of other sciences seems to be inevitable. We saw one example of this with the pandemic: if we have to think about who meets whom, then understanding network properties is crucial for being able to say anything about the spread of a pandemic. Networks are also needed, of course, in brain research and ecological research, for example. So, with or without limits, the study of very large graphs and their properties, the problem of how to model and study them, is very important. And these are all very difficult questions, so I think that the study of large graphs is an exciting new area.

I completely agree with you, of course. Thank you so much for this very interesting and inspiring interview!

4

Rianne de Heide: From Statistics to Music, with an Open Conversation on Mental Health

Rianne de Heide (Fig. 4.1) is an Assistant Professor in the Department of Applied Mathematics at the University of Twente. She was born in Rotterdam, The Netherlands, in 1989. She obtained a bachelor's degree in classical music (horn) from the Prins Claus Conservatoire in Groningen (2012), a bachelor's degree in mathematics from the University of Groningen (2013), a master's degree in classical music (horn) from the Royal Conservatoire in The Hague (2014), and a master's degree in mathematics from Leiden University (2016). In 2021, Rianne obtained her PhD in mathematics at Leiden University and at the Centrum Wiskunde & Informatica (CWI) in Amsterdam. She then moved to Magdeburg, Germany, for a postdoc at the Otto von Guericke Universität, but she was soon awarded an NWO Rubicon grant and therefore moved to Villeneuve d'Ascq, France, to work at INRIA Lille. From 2022 to 2024, she was an Assistant Professor at VU Amsterdam, and in 2023 she was awarded a VENI grant from the Dutch Research Council. In 2024, she moved to the University of Twente.

Fig. 4.1 Rianne de Heide. *Photo: Annabel Jeuring*

Raffaella Mulas interviewed her in August 2023.

Thank you, Rianne, for taking the time to meet me. Usually, when I interview someone, I start by asking them to tell me something about the beginning of their life and career, but with you, I would like to start from the latest news. You will soon go on maternity leave, and you will then come back as a VENI grant recipient. Can you tell me something about your VENI?

Yes! My proposal is about hypothesis testing, which is what I mainly work on and it's what you perform, for example, when you want to understand whether a pill is better than a placebo. I'm going to introduce a new theory for hypothesis testing in the field of multiple testing. When you do many, many hypothesis tests, like 100,000 or 400,000 hypothesis tests at the same time, e-values offer greater flexibility compared to the classical methods that are used now.

Great! And this is about your future research. But I would like to ask you something about your past research as well, because despite your very young academic age, the VENI is only one of the several grants that you were awarded. You also received the Van Zwet Award 2022 from the Netherlands Society for Statistics and Operations Research (VVSOR) for your PhD dissertation. So it's not too early to ask you this question: What is your past research achievement that you are most proud of?

Finding the connection between group theory and hypothesis tests. There was a big open problem on which also my PhD supervisor, Peter Grünwald, had great breakthroughs, which is how to make hypothesis tests that you can do sequentially in time for problems with a composite null. This led to our new theory for hypothesis testing with e-values. We formulated an optimality criterion for those tests, and then, I found a connection between group theory and such optimal tests. This led to several papers, and coming January we are invited to London to present our first pioneering paper about our new theory for hypothesis testing at the discussion meeting of the Royal Statistical Society. It is considered a significant recognition to be invited to present a paper there.

This is amazing! When and why did you decide to study both mathematics and music?

After high school, I actually started to study physics and conducting. But I conducted some youth orchestras and I didn't find it very pleasurable, so after one year I decided to play an instrument instead. I applied to studying horn at the Prins Claus Conservatoire, and I auditioned and got in. Around the same time, I was studying physics and I really liked the theoretical subjects like relativity and quantum mechanics, but I hated the practical ones. And when I followed my first algebra course I thought, okay, I should switch to mathematics.

What do music and mathematics have in common for you?

A lot of people think that there are a lot of things in common between music and mathematics, but I don't think that this is really the case. For me, the thing in common is the great pleasure that I take in doing both. The feeling when you finally have the proof of some mathematical result is comparable to the feeling when you play a nice solo in an orchestra in a concert.

Interesting! And how is music a part of your life today?

Today, I'm singing in the VU-Kamerkoor, because when I became a professional horn player I needed a new hobby, so I started taking singing lessons! I'm very much enjoying that, especially since my best friends sing there as well. I have some side projects in other choirs and even some professional choirs which I really like. I also still play the horn in a semi-professional orchestra.

What are the things you like the most in academia?

I love the freedom that I have to conduct my own research. I love the challenge that mathematics gives me, and I love the fact that I'm never done with learning new things. I also like to cooperate with many people, and I really like the type of people that one finds in mathematics and in academia. When I was at school and even during my studies, I always felt like the odd one. But here I can really be myself and I still feel appreciated by everybody because of who I am. I have the same feeling with my choir.

This is beautiful! Now, when I was preparing for this interview, I did a bit of internet stalking (as usual), and I found a picture of you in what appears to be a TV quiz show—is this correct?

Yes! In 2018, during my PhD, I took part in a quiz on Dutch television called Per Seconde Wijzer, in the science category. I got some mathematics questions, but also questions from other areas like biology, philosophy, and even things like to identify the breed of dog from a picture of a puppy. It was super stressful because of the time pressure, but it was also super nice, and I won a nice amount of money!

This sounds amazing! Besides mathematics and music, what makes you happy?

My dog and my partner, and I'm looking forward to adding a kid to the list soon! I also enjoy spending time with friends and doing sports. I like long distance running and cycling. And now that I'm pregnant and I can't run, I started swimming.

Nice! Before this interview, you mentioned that you would like to take this opportunity to open up about something important: What would you like to share?

Yes. Since I started my studies, I've been struggling with a mental illness, and this made several periods in my life very hard. I struggled in particular during my studies, during the postdoc phase because I had to move abroad and leave friends and family behind, and during the COVID lockdowns. Now I would like to talk about it because I think that mental illness should be normalized and treated like any physical illness. In the same way a friend of mine is sick sometimes because she has diabetes, I am also sick sometimes because of this. And I also want to talk about it to show that it is possible to have this issue and still have a successful career as an academic.

It has a huge impact on my life, and I live in a very structured way: I have a very healthy lifestyle, I do a lot of sports, I always go to bed early. I do all these things to prevent getting sick again. And I still get sick, but now I know that it always goes away at some point, and this gives me strength. When my brain is fully functioning again, I can do all the things that you see on my CV: I can write nice papers, I can obtain grants.

I also want to say: I don't want to share all the details on the internet, but I do want to encourage people to come to me and talk to me about it.

Thank you for being so open and sharing this. Mental health is such an important topic, and not only you are talking about it, but you are also giving a very positive and inspiring example. Did you already open up at work in the past?

I actually really opened up for the first time, last December, with my mentor here at VU, and it was such a relief! Before, I shared some bits with my PhD supervisor, but not all, because I was always scared that nobody would hire me for my next job. But I don't have this fear anymore, because people know my value. I get all these prizes and grants, and this shows to the outside world that I'm worth my value as a researcher, I think. I feel much more confident now, and together with the stability that this job gives me compared to a postdoc, this has also improved my mental health a lot!

This is very nice. If you could go back in time, would you open up more?

Yes. I think that things would have been better during my PhD, if I had shared more. My supervisor was super supportive, and in hindsight, I think I should have shared more with him.

What advice would you give to a PhD student who is struggling with similar issues?

It depends very much on the situation. Not everybody should just share everything, because there's still so much prejudice and stigma around these issues. I think it's safe to not disclose too much in general. But if you have a nice supervisor who you think is supportive, then maybe it's good to share these things with them. And in general, it's good to open up at least with someone. It can also be a friend. I actually kept this as a secret with all of my friends for many years, but I then found it super helpful to open up with them. And especially during periods when I was feeling bad, they have been super supportive. In fact, opening up to my friends and seeing that they supported me instead of rejecting me, was a real turn in my life!

This is beautiful. You're giving so many beautiful messages here. Thank you so much!

5

Science, Theater, and Intellectual Curiosity with Max Planck Institute Founder Jürgen Jost

Jürgen Jost (Fig. 5.1) is an Emeritus Professor of Mathematics and Emeritus Director of the Max Planck Institute for Mathematics in the Sciences (MPI-MiS) in Leipzig, Germany, and an External Professor at the Santa Fe Institute in New Mexico, USA. He was born in Münster, Germany, in 1956. He studied mathematics, physics, economics and philosophy at the University of Bonn from 1975 to 1980, and in 1980 he also completed his PhD in mathematics at the same university. He has held various postdoctoral and visiting positions at IAS Princeton, UC San Diego, ANU Canberra, MSRI Berkeley, Harvard, ETH Zürich and IHES Paris. From 1984 to 1996 he was Professor of Mathematics at the Ruhr University Bochum, and in 1996 he moved to Leipzig, where together with Eberhard Zeidler and Stefan Müller, he founded the MPI-MiS. To date, Jürgen Jost has written more than 600 research articles and more than 20 books, spanning many different areas of mathematics and applied sciences, as well as philosophy and history of science.

Fig. 5.1 Jürgen Jost. Photo: Cressandra Thibodeaux / Santa Fe Institute

Raffaella Mulas (who is one of Jürgen Jost's former PhD students) interviewed him in November 2023.

Thank you, Jürgen, for agreeing to do this interview! I would like to start with an unusual question. Of course, you know that the Erdős number of a researcher is the distance from Paul Erdős in terms of research collaborations. And you might also know that the Bacon number of an actor is the distance from Kevin Bacon in terms of acting collaborations. One then defines the Erdős–Bacon number as the sum of a person's Erdős number and Bacon number. Do you know what your Erdős—Bacon number is?

Well, I think that my Erdős number is three. I don't know what my Bacon number is, but it's probably low because I recently acted together with Hanns Zischler, who is a well-known German actor.

This is exactly what I wanted to talk about! I have checked it, and because of your collaboration with

Hanns Zischler, also your Bacon number is three. So you have Erdős–Bacon number six, which is exceptionally low.

Haha, thank you! Very interesting information that I am learning from this interview!

Can you tell me something about this theater play?

Well, it originated when Hanns Zischler at the Academy of Science and Literature at Mainz, of which we are both members, asked me whether I would be willing to perform together with him in a play that he had seen and translated into German. The play was about how John von Neumann, after his death, had a meeting with God and asked many questions but ended up arguing because God did not provide satisfactory answers. I said yes to Hanns, but when I asked him to send me the script and looked at it, I found that while the basic idea was very good, the details needed a lot of rewriting. So, I kept the original idea of the plot, but otherwise I completely rewrote the piece. It turned out that Hanns liked my version, and we decided to perform it! I had the role of John von Neumann, and he had the role of God, which, of course, was completely appropriate for him. We first had a practice session in Leipzig, and then last June we performed the play at the Academy of Sciences and Literature, in a public event in Mainz. And it was quite well received, somewhat to my surprise because I'm not a professional actor, whereas Hanns, of course, is one of the best German actors.

I am not at all surprised that it was well received, and I hope that you will perform again in the future! Can you tell me something about your PhD? On paper, it looks like it was two-months long, but you explained to me that you actually worked on your PhD project while you were still an undergraduate student.

Yes, exactly. Well, at some point, I was undecided between pursuing a career in mathematics or economics, and I had an offer for a PhD position from a very well-established economics professor in Bonn—Wilhelm Krelle. But then I also talked to Stefan Hildebrandt, and he gave me some problems for both a diploma thesis (which corresponds to a master's thesis today) and a subsequent PhD thesis. These problems were in the field of geometric analysis, and they involved applying methods from partial differential equations to geometric problems, which at the time I found very interesting. I first looked at the diploma problem that he gave to me, but I figured out pretty quickly that this wouldn't work, and I explained that to him. He said: "Why don't you start working on the PhD topic already? This is a very interesting problem, and if you solve it, it will be quite well received by the community!". So, I looked at it, while I was still an undergraduate student, and I found out that it was easier than what he had anticipated, so I could solve it relatively quickly and I submitted it as a PhD thesis two months after submitting my diploma thesis. In fact, it could even have been slightly earlier, but in between, he was away for a sabbatical in the United States, and at that time, communication was by ordinary mail, so it was not so quick.

Now, once I was having dinner with the research group of Bernd Sturmfels, when Bernd said: "You probably all know that the Max Planck Society has three sections: the Biology and Medicine Section, the Chemistry, Physics and Technology Section, and the Human Sciences Section. But what you might not know is that, among the 25,000 people who work at the Max Planck Society, only one person is a member of all of its sections. This person is Jürgen Jost!". Another time, your collaborator Marius Gardt from the European Central Bank said to me: "Jürgen has such a deep knowledge and understanding of economics, that when I talk to him, I cannot believe that he is a mathematician!". And countless times I have

heard mathematicians talking about how unbelievable it is that you have made important contributions in so many different areas of mathematics. So, my next question is: How do you do this? How can one person be so productive in so many different subjects and areas?

This is, of course, first of all, a very kind way of putting it. But over the course of my life, I have learned to systematically explore connections and relations between different fields, and that is probably my strength. I can put things that I learned in some field into perspective and translate it into a mathematical question or relate it to a question in some other field. And so, of course, this makes it relatively easy for me to work in different fields. I can analyze the conceptual structure of one field and relate it to formal structures that I may know from mathematics or from other areas.

This is very interesting! A related question is: What are you not good at?

Well, I have no talent for music, for instance! I took flute lessons when I was young, but without much success.

In mathematics, what is your creative process? How do your ideas take shape?

I don't know whether there is any general answer to that, because it depends on the particular problem. But of course, since I've explored many different fields, I know some techniques or methods from different areas that I can then bring to bear on a problem. For example, at the beginning of my career I was combining geometry and analysis, and in particular nonlinear partial differential equations and the calculus of variations. But I understood more geometry than most of the people working in analysis, and in contrast to many people who were then working in geometry, I was not afraid of applying different difficult methods from nonlinear analysis, from partial differential equations and the calculus of variations. And while other people were very good in quickly doing complicated estimates with partial differential equations, I rather tried to use some geometric imagination to understand what was behind it. So, combining different fields can make you productive because then you have access to different concepts and methods. I did this also for example when, later on, I moved to discrete mathematics. Exploring analogies between Riemannian geometry and graph theory helped me a lot to understand and address questions in discrete mathematics from a different perspective than the people who are usually working in these fields.

Is there a mathematical result you are most proud of?

I think that a very good mathematical result of mine is the existence and the control of so-called generalized harmonic maps, or equilibrium maps, between metric spaces when the target has generalized non-positive curvature. This result opened many directions, and it was also useful for looking at certain problems in algebraic geometry when it came to representations of fundamental groups in fields of non-zero characteristic. I worked on this in the 1990s with Kang Zuo, who is a very good algebraic geometer.

You will retire in 2024. What will change for you, and what will remain unchanged?

Of course, one thing that will change is that I will no longer have the possibility of having a large group of students and postdocs. Usually, I have between 30 and 40 people working with me in Leipzig, if I count students, postdocs, regular and long-term visitors. So, the group will become much smaller, but I will still have a number of PhD students and also some postdocs for some time. And I just plan to continue to work as a scientist. I have many ongoing research projects and books that I want to complete, and of course, I'm open for new scientific problems with new collaborators.

So, it sounds like, apart from the size of your group, not much will change. This is nice! What advice would you give young researchers?

It's always good to try to go into new directions that have not yet been explored so much but offer exciting questions and problems. So, my advice is: be open to new directions. Do not just follow others and try to slightly improve the results that some great minds before you have achieved already. Try to find your own way!

This is great advice! Besides research, what makes you happy?

Clearly, my family is a great source of happiness!

Besides that, I am an outdoor person and like hiking and biking, and I also have many, many cultural interests. I read a lot of literature for instance. I have many friends who are writers or poets, and talking to them is of course very stimulating.

Thank you so much for this very inspiring interview!

6

EMS Prize Winner Cristiana De Filippis' Perspective on Research and Life's Balance

Cristiana De Filippis (Fig. 6.1) is an Associate Professor in the Department of Mathematics at the University of Parma. Born in Bari, Italy, in 1992, Cristiana obtained her bachelor's degree in mathematics at the University of Turin and her master's degree at the University of Milan-Bicocca. Then, she completed her PhD at the University of Oxford. Following her PhD, Cristiana held a postdoctoral position at the University of Turin from 2020 to 2021, after which she moved to the University of Parma.

Throughout her career, Cristiana has been awarded several prestigious recognitions, including appearing on Forbes' list of the 100 most successful Italian women in 2023 and being one of the top-cited mathematicians for her PhD year (2020). In 2024, she received the Bartolozzi Prize by the Italian Mathematical Union. In the same year, during the 9th European Congress of Mathematics, Cristiana was awarded the EMS Prize, which is considered the most prestigious mathematics prize in Europe and is referred to as the "Fields Medal's little sister."

Fig. 6.1 Cristiana De Filippis and Anakin. *Picture courtesy of Cristiana De Filippis*

Raffaella Mulas first met Cristiana De Filippis through the inaugural cohort of the European Mathematical Society Young Academy (EMYA), where they both serve as elected members, and interviewed her in July 2024.

Cristiana, congratulations on receiving the EMS prize! It is a great honor for me to interview you.

Many thanks to you Raffaella for the interview, the honor is mine!

I would like to start from the beginning, by asking you how you discovered your passion for mathematics, and in particular how old you were.

I discovered my passion for mathematics right after primary school: I realized that studying was pleasant and exercises/problems were easy, so I also deepened some aspects of the subject by myself.

Well, that turned out to work out very well! Now, you were awarded the EMS prize for your "outstanding contributions to elliptic regularity, in particular Schauder estimates for nonuniformly elliptic equations and non-

differentiable variational integrals, and minima of quasi-convex integrals." How would you explain this to someone who thinks that "elliptic regularity" is a type of workout routine?

"Workout routine" is very nice, I will use it somewhere! Let's say that an "elliptic PDE (Partial Differential Equation)" is an equation driven by a differential operator that is a (possibly degenerate) generalization of the Laplacian. Investigating the regularity of solutions means finding out how "good" they are. Two basic questions may be for instance, "Can their graph be drawn with a continuous line that has no holes?" or "How many derivatives solutions have?". To answer, it is in general necessary to design delicate iteration schemes aimed at showing that certain features of solutions improve at each step, in a controlled way. This in general requires a deep understanding of the problem, very fine techniques as well as a bit of luck.

You have all three! What is your creative process? How do your mathematical ideas take shape?

My creative process is rather visual, I tend to imagine what are the right paths towards the outcome and the main obstructions, leaving details aside. Sometimes it works, others a super small parameter appears in the wrong place and nullifies months of work.

Does this mean that also an EMS Prize winner can have difficulties?

Of course! I feel stuck for so many things in my research, and what comes out is just the tip of the iceberg.

Can you share a specific moment or breakthrough in your research that was particularly exciting for you?

When I obtained the first set of Schauder estimates for nonuniformly elliptic PDE, in collaboration with Giuseppe Mingione. This was probably the first "big" problem I've heard about at the very beginning of my PhD, with the warmest recommendation of not wasting my time working on it, as nothing would have come out.

Great that you didn't listen! Cristiana, I looked at your CV and I am impressed by the breadth and depth of your activities and achievements. How do you manage your time to balance research, teaching, and the various other professional commitments that you have?

You are very generous, thank you! I focus the majority of my time on research, and try to optimize teaching in such a way that the efforts required remain reasonable.

What advice would you give to younger researchers who are just starting their journey in mathematics?

My advice would be to always be curious, willing to study, and not be afraid of hard work and failures.

Great advice! Since we are friends on social media, I have to ask you: Do you manage your accounts, or have your cat taken over as the real influencer behind the scenes?

That's right, you spotted the real social network mastermind, the king of perfect shots, the nightmare of the average influencer, the terrible cat Anakin!

I knew it! Who have been your greatest supporters throughout your career?

My horse and my cat, of course! And my friends (colleagues) at the University of Parma. We made a very strong and supportive group, so it's natural in such a nice environment to try to do our best.

This sounds wonderful. Besides research, your friends, and your beloved animals, what makes you happy?

I like running and doing sports in general!

And what are you looking forward to in the future?

I look forward to continuously improving my work and the results I obtain. I also hope to be able to build a research group to share ideas, approaches and perspectives and, above all, to learn from other researchers with different backgrounds from mine.

I'm sure it will happen!

7

Renee Hoekzema: Blending Earth Sciences, Topology, and Artistic Expression

Renee Hoekzema (Fig. 7.1) is an Assistant Professor in the Department of Mathematics at VU Amsterdam. She was born in Utrecht, The Netherlands, in 1987. In 2009, she earned two bachelor's degrees in mathematics and physics/astronomy, with a minor in biogeology, from Utrecht University. She then furthered her education at the same university, obtaining two master's degrees in mathematics and theoretical physics in 2012. Following her studies in Utrecht, Renee pursued doctoral studies at Oxford University, completing a PhD in earth sciences in 2015 and a PhD in mathematics in 2018. Her academic journey also includes postdoctoral appointments at the University of Copenhagen, at the Max Planck Institute for Mathematics in Bonn, and back at the University of Oxford. In 2022, Renee was awarded a VENI grant from the Dutch Research Council, and she joined VU Amsterdam.

Raffaella Mulas interviewed her in April 2024.

Thank you, Renee, for taking the time to meet me. I would like to start by talking about your challenging,

Fig. 7.1 Portrait of Renee Hoekzema made by her mother, Anneke Kerkhof

fascinating and very unique academic path. What drove you to pursue such diverse fields of study?

Thanks for interviewing me!

I think I always had a very broad interest. When I started university I really wanted to study all three subjects: physics, maths and earth science. I might have finished three bachelors and three masters degrees, but someone recommended that I should just start doing research instead. So, after my maths & physics masters, I started a PhD in paleontology. The research was fun, but I also missed doing more maths. In fact, I had a dream that I was a mathematician and when I woke up I knew that that's what I wanted to do! So I finished the paleontology PhD and started a second one in algebraic topology with Ulrike Tillmann, who was very supportive.

In hindsight, the lack of female role models had contributed to me not choosing to continue with maths before. Meeting amazing female mathematicians at Oxford such as Ulrike Tillmann and Francis Kirwan changed my point of view. As a mathematician I now do research in maths, physics, biology and even paleontology, so in the end I'm combining all of my interests!

You made very exceptional choices! And I understand your point about the importance of having female role models. I find it impressive that you were able to merge your different expertises in mathematics, physics, paleontology and biology. Not only do you have a deep knowledge of each of these subjects, but you also connect them in your research, and you did so in your VENI grant proposal as well. Can you tell me something about it?

As part of my maths PhD I studied "shapes inside shapes" and their evolution through time (in maths language: manifolds with submanifolds and cobordisms on them). When I started writing the VENI, someone recommended that I stay close to my heart, and rather than thinking about what would please a committee, write the proposal about what I really wanted to do. Now the cobordisms theory I was studying had close links to certain theories in mathematical physics called topological quantum field theories, so there was a clear link there. On the other hand, during my masters I had done an amazing course on vertebrate paleontology for which I wrote an essay titled "Evolution of vertebrates through the eyes of parasitic flatworms" (which I personally thought was hilarious because flatworms don't really have eyes—I don't think anyone got the joke). Back then I was already curious about whether one could make a mathematical model to study the phylogeny (evolutionary descent) of a group of hosts together with the evolution of the parasites living on them (and co-evolution more generally). I realized that from the point of view of maths, this is also a shape within a shape evolving through time (namely a tree inside a tree), and therefore connected to the maths/physics projects that I had been thinking about. Putting these all together, I wrote a proposal called "Shapes inside shapes in mathematics, physics, and biology".

Amazing! There is also another subject that plays a role in your VENI grant. At the end of your proposal, you talk about collaborative projects with artists along the theme of "shapes inside shapes"...

I grew up in a crossover art/science family, so art has always been close to my heart. I love drawing and painting and I also like to sing. I'm currently setting up a collaboration with a pair of installation/theater artists that aims to capture the experience of understanding something in maths, without explicitly involving maths itself, but with a flavor of shapes inside shapes. It's a very exciting process!

Beautiful! Now, I need to ask you: Renee, what are you not good at?

Haha, there's plenty of things I'm not good at! I have a terrible memory for example, and I'm easily distracted.

But this doesn't stop you from doing anything! You are also a tireless traveler. You travel a lot for conferences and research visits, but also for pleasure. What has been your most beautiful trip so far?

Last year my partner and I made a tour of South East Asia which was amazing. We spent a few days on a Thai island where the beach was perfectly white and the sea was turquoise blue and crystal clear. You could walk into the water and immediately see corals and seahorses. I didn't know a place like that existed!

I also feel very fortunate being able to make beautiful trips for work. Recently I had a conference in Kyoto for example, which was also stunning.

And the most adventurous one?

I made a solo trip to Costa Rica at some point which was really spectacular! The Costa Rican rainforest is one of the most biodiverse places on Earth, and I found myself thinking about how every square meter contained at least one animal that could kill you (or at least seriously poison you). From the cute poison dart frogs to the vipers, scorpions and Brazilian wandering spiders—one of the most deadly spiders in the world, of which I saw quite a few. At some point at the end of a night tour of the jungle, my phone had died after taking many pictures of snakes and giant insects. I didn't have a torch and I had to walk back through the jungle alone in the dark. That was probably one of the most adventurous things I've ever done.

Wow! Besides traveling, art and research, what makes you happy?

Spending time with my friends, family and partner, and our baby on the way!

Nice! I love the portrait of you that your mother, Anneke Kerkhof, painted and that you have on your website. May I use it for this article?

Of course! My mom's artwork has always been the decor of my life and she's a great source of inspiration for me, so I was honored when she decided to make a portrait of me.

You are going to be a great source of inspiration for your baby! Thank you for all the beautiful things that you have shared.